A Journey towards Eco-friendly Grain Storage

A Grain Saved is a Grain Produced

Dr. S. Mohan

Retired Professor of Agricultural Entomology
Tamil Nadu Agricultural University
Coimbatore (India)
Email :sarmamohan@hotmail.com
Web: www.mohantrap.com

SCIENTIFIC
PUBLISHERS

Published by

SCIENTIFIC PUBLISHERS (INDIA)

5 A, New Pali Road
P.O. Box 91
Jodhpur - 342 001 INDIA
E-mail: info@scientificpub.com
Website: http://www.scientificpub.com

First Edition, March 2019

ISBN: 978-93-89061-53-6 (Paperback)
978-93-89061-54-3 (E-ISBN)

Printed in India

Dedicated to

Tamil Nadu
Agricultural University

Paul Fields
Research Scientist
Agriculture and Agri-Food Canada
Winnipeg, Canada

Foreword

Why is the sky blue? Why is my hair straight? Why does the sun go away at night? Why can't I fly like a bird? “Why this, why that…” the stage that every young child goes through and every parent struggles to answer the deluge of questions. Children are naturally curious about the world.

As noted by Neil deGrasse Tyson, the renowned American astrophysicist, author, and science communicator; “Don't get in the way of children who find it natural and obvious to explore the world around them - even if it means they make a mess of your kitchen or living room. It's all about your perspective on these things. Let them play. When you do, the kids do not have to be reintroduced to ways of questioning nature, and the task of promoting science would be a trivial exercise.”

Dr. S. Mohan came to my laboratory in Winnipeg Canada at Agriculture and Agri-Food Canada, a senior scientist, but with his child-like curiosity intact. I had many lively conversations with Dr. S. Mohan over the years discussing the behaviour and biology of insects and in particular those insects that are fond of eating the seeds that farmers have worked so hard to produce.

I hope you enjoy reading this book as much I have, and if you have the chance, take on a career as a scientist.

Paul Fields

Contents

Chapter 1

A Grain Journey Begins

We live in an era of food safety. Hence it is imperative that grains and seeds produced by the farmer through backbreaking toil and arduous labour be saved. Moreover, in a world where millions go to sleep on an empty stomach, wastage of food grains even before they reach the market, due to ignorance, primitive technology and lack of advocacy and education among farmers is a matter of great concern. It is in this context that technology gains significance. Technology not only helps save money, but also saves food grains and this saving must begin from home. "Food safety and clean food" is the mantra being followed by the world today and we cannot understate the importance of the use of technology for advancement of our farming methods.

When we inculcate the culture of saving grains and seeds in every home and farm, we start benefitting immensely and would be ready to use any technology to prevent wastage of grains due to destruction by insects, especially if it is user friendly as well as eco-friendly. To amass a huge savings of grains that would benefit the entire nation and the farmer in particular, I have taken conscientious efforts to document my knowledge gained as an agricultural scientist as well as my insightful experiences, in the form of a book that will highlight the importance of equipping the farmers with better methods of grain storage apart from use of technology to upgrade farming techniques. This book will be enlightening and offer deep insights into the genesis, development and use of a technology developed by me for preventing wastage of millions of tonnes of food grains, benefitting not only farmers and institutions that render services to agriculture, but also readers. This book will empower readers to educate those involved in agriculture to follow the practice of saving in every which way for larger benefits.

Is it easy to develop a technology and reach it to users?

No way. Not only developing a technique has its own challenges, but it is also a Herculean task to motivate, persuade and convince farmers to adopt new technologies and agricultural methods. Unless a "seeing is believing" approach is targeted toward the users, fostering a change in the outlook of farmers will be a tough task; but it is never impossible. Here is the success story of my life long experience (1980-2018) as a research scientist in Tamil Nadu Agricultural University (TNAU), Tamil Nadu, India which paves way for much optimism in this regard. This book is the story of my long and progressive journey with a vision to reach 20 to 30 crores farmers/users across the world using the technology I have created and developed thoughtfully and purposefully.

What is the present agricultural practice?

Before understanding and accepting new technologies, it is important to be informed about the present trends and practices in agriculture especially that of food grain storage.

Soon after harvest, farmers dry the produce below 10% MC (moisture content) and store the dried produce in gunny bags and bins. Interestingly, the harvested produce contains insects in different life stages such as the egg, larvae or pupae (which are inside the grain) and which are carried over from field. It is well proved by researchers that, field-laden damage is <1% but if this is not controlled, a total loss could possibly happen within 2–3 months. It is obvious that in field-laden infestation, adult insects generally emerge between 15–20 days after the storage and start mating, laying eggs and multiplying into large numbers. These insects have a high reproductive rate.

The traditional practice to ward off insect-pests adopted by farmers is through sun-drying. The farmers generally sun-dry the produce a month after storing the food grains. By this time, the eggs of the insects hatch and enter the grains and become concealed in most of the cases. Sun-drying can destroy and kill only the eggs and not the larvae or pupae, which are found concealed inside the grains. Hence, there is an urgent need for timely detection of infestation carried over from the field, during storage.

Here comes the need for use of my bin (or) bag storage device, both for the control and the detection of 'Zero Tolerance' limit as prescribed by many countries for many commodities (with reference to insects) for export under WTO treaty.

According to the Prevention of Food Adulteration Act, Government of India (GOI), the amount of weevilled grain (insect damage) (by count) shall not be more than 10 per cent (or) Uric acid content arising as a result of insect damage shall not exceed 10 milligram per 100 gram, whichever is lower.

Annually, on an average 2-3% of stored food grains is lost due to insect pests, which accounts for nearly Rs. 1300 crores.

A pair of insects, through three generations a year, has the biotic potential to reproduce 1,500,000 off springs, which will consume 1,500,000 kernel of rice (Amounting to 30 Kg of rice).

The above information is sure elicit the concern of the readers and throw some light about the problems I tried to address to save the food grains stored that are the results of hard labour of our farmers.

My journey as a researcher had begun when I tried to address the above issues during 1980s. When I started my research career in 1980's, it was an era of Integrated Pest Management (IPM), where *Helicoverpa* spp., on cotton, white fly on cotton, and Tungro virus in rice were given focus. These were considered to be burning issues in agriculture and hence no new thinking or work evolved with regard to "Stored Product Insect Management."

But my focus was different from that of the others in the field. I was constantly tormented by an oft-repeated question as to what the consequences would be if the grains or seeds that were produced through back-breaking labour by farmers were not stored properly at home / farm / warehouse.

This disquiet urged me to launch into a research on the available literature on developing eco-friendly / new strategies for stored grains insect management. Unlike the present times, back then, we did not have a large-scale resource base as well as access to universal knowledge and information through the Internet etc. TNAU library was the only source of knowledge and only hard copies of journals were available that came to my rescue.

The turning point in 1984-1985

I was fortunate to go through the special issues of Canadian Entomologist Journal on the detection of stored product insects by eminent foreign authors, that included a reference paper by Loschiavo S.R, Atkinson J.M 1967. A trap for detection and recovery of insect in stored grain, *Canadian Entomologist 99 (11): 1160-1163*

I was impressed by the research paper of Loschiavo, S.R. and decided to adopt a similar approach for India, so that one day the subject of stored grain insect monitoring and management would be understood throughout the nation.

This led to a series of surprising developments in my journey as an agricultural scientist.

A gratifying time with Dr. S.R. Loschiavo

I happened to be with S.R. Loschiavo in his home for tea in 2000 during my training at Cereal Research Centre (CRC), Winnipeg, Manitoba, Canada, at the same laboratory where Dr. S.R. Loschiavo worked and developed the first monitoring device. This was a providential occurrence.

I was instrumental in transforming our country into one of the world leaders in stored grain insect monitoring science as evinced by the invitation to the **11th International Working Conference on Stored Product Protection, November 24-28, 2014, Chiang Mai, Thailand.** I was the first Indian to be invited as a key-note speaker in this field, which was really overwhelming.

Opportunity rocked in Chiang Mai, Thailand .....
An overwhelming moment in my life.

How did the turning point unveil itself? How did I reap success for my endeavours in this field and what was the reason behind this success? The success quotient is not only for the farmers across India, but also for farmers in several African countries who witnessed my technology and believed in it. This technology that I have developed has now been adopted extensively across the globe.

Scaling the pinnacle of success

The first device, the modified version and a form suitable for our country was released in 1993, which I developed as a part-time Ph.D. student of TNAU working in ICAR (Indian Council of Agricultural Research) scheme on Harvest and Post-harvest technology.

To my utter amazement, this invention fetched me a cash award from the Tamil Nadu Government in 1993 and the device was named after me as **"Mohan Trap"** by the Tamil Nadu Government through a Government Order.

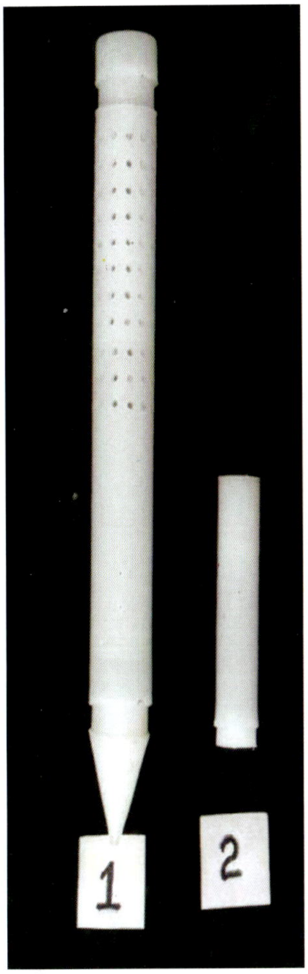

Plastic Preliminary Model Trap (Mohan trap)

Awarded a cash of 5000 INR

The Tamil Nadu Government did not stop with this. The Government was very much interested in taking this technology to the farmers, entrusting the job to the then Tamil Nadu Agro Industries, by providing 1,00,000 INR as assistance for manufacture and distribution of the Mohan trap.

This encouraging move by the government resulted in the spread of the trap technology to around 500-1000 farmers. Thus the first seed toward safe grain storage by adopting my technology was sown by the Tamil Nadu Government by extending their support to TNAU trap. Following this, I made continuous efforts to popularize the use of Mohan trap through my articles in eminent publications that became popular and through other extension mechanisms for intensive and extensive propaganda.

As the years moved on, yet another significant development took place in 2002. Tamil Nadu Agricultural University signed an agreement with KSNM marketing, Coimbatore, Tamil Nadu, India to manufacture and market TNAU stored grain insect trap. Thus the commercial model snared the market extensively.

Commercial Model

Commercial Launching of TNAU Probe Trap by TNAU in 2002

About the Trap

1. Insects love "AIR" and move towards air. This behavior of the insect is exploited in this technology.
2. The insect trap has to be kept in the grain like rice, wheat etc., vertically with the white plastic cone downside as shown in the figure.
3. The top red cap must be with the level of the grain. Insects will move towards air in the main tube and enter through the hole.
4. Once the insect enters the hole it falls down into the detachable white cone at the bottom. Then there is no way to escape and the insects are trapped forever.
5. The white detachable cone can be unscrewed once in a week and the insects can be destroyed.

This became a historical event as it lead to the spread of about 5 lakhs TNAU insect trap (0.5 Million) to farmers and end users across the country.

The success of the trap was due to the feedback from farmers and users under a massive programme - TNAU–McGill-CIDA Food Security Project. How did the trap technology spread to 5 lakh users (0.5 Million)? The photo feature which follows will bring to light the modalities of the successful reach of the trap.

TNAU Probe Trap (Mohan Trap) Technology Spread across World

Coimbatore, Tamil Nadu, India.

From **Pallapalayam**, *Village, Tamil Nadu, India*

To

France

Next My aim is to take the Trap

Proud moments receiving Outstanding Extension Scientists Award in 2010 by ICAR.

And this paved the way for yet another development in the story of my trap.

Any guesses? Well, here goes the story

Success story of a Farmer's Innovativeness based on TNAU Probe trap

A farmer in a village in Tamil Nadu by name Mr. V.K.V. Ravichandran, 28, Nallamangudi, Agraharam, Nannilam, Tamil Nadu, India - 610 105, inspired by the TNAU probe trap, made his own model with waste water bottles and used them in his paddy storage structure. This success story was published in a Tamil magazine "Tamil Vivasaya Ulagam", P.33-34, May 2008.

Mr. V.K.V. Ravichandran, Tamil Nadu with his creative excellence

Chapter 2

Transforming into an Inventor

More often than not, I start with the literature to work on. So far you have read about the impact of TNAU probe trap and how it reached people across India. My efforts to spread the trap technology were ongoing.

After developing a form of trap suitable for our country based on literature,

I launched into an introspection about what my next move would be. Practically a scientist gets the title "Inventor" from a society only if he invents a new technology or a product.

Constant brain racking led to the emergence of a new idea to make a breakthrough and emerge as an inventor, which would help me connect the entire world with my much revered TNAU and in turn with my state and the country as a whole.

The breakthrough happened with my invention of the TNAU Insect Removal Bin, fetching me a coveted honour and recognition and establishing me as an inventor in 2002.

I won the invention ward for my TNAU invention 'Automatic Insect Removal Storage Bin'.

National Technology Day Invention Award, 2002, Government of India

I had never imagined this would happen. But the fact is imagination and inspiration that result in diligent research always lead to inventions. So what is this bin all about? It is now time for me to outline the details of this innovation to satiate your curiosity.

TNAU insect removal bin is a device which can remove insects automatically. The structure consists of 4 major parts namely, outer container, inner perforated container, collection vessel and the lid. The model exploits the 'wandering behaviour' of stored product insects as well as the movement of these insects towards well aerated regions. The grains are stored in the specially designed inner perforated container. The space between the inner and outer containers provides adequate aeration for the insects. The insects, while wandering, enter the perforation to reach the aerated part and while doing so, slip off and fall into the collection vessel through a pitfall mechanism provided in the collection vessel. To collect the trapped insects, as and when they emerge from grains, perforated (2 mm) rods are fixed in the inner container.

The container serves as a storing bin for rice, wheat, broken pulses, coriander seeds, etc. The insects such as rice weevil, lesser grain borer, red flour beetle, saw toothed beetle, which are commonly found attacking stored grains are removed automatically by storing grains in the container. Within a very short span of 10 days a majority of the insects (more than 90 per cent) can be removed from the grains. The containers are available in 2 kg, 5 kg, 25 kg, 100 kg and 500 kg capacities.

Efficiency

Grains (paddy and sorghum) stored in Automatic Insect Removal Bin (100 kg and 500 kg) recorded only 1 – 4% damage by insects, compared to 33 to 65% damage on grain storage in ordinary bins after 10 months of storage. The population of insects (*Rhyzopertha dominica., Sitophilus oryzae.,)* ranged between 0 – 2 / kg in grain stored in 100 kg automatic insect removal bin compared to 5–191 / kg in ordinary bin after 10 months of storage.

How did Automatic Insect Removal Bin Happen?

A strong will to find a comprehensive solution to minimize grain loss coupled with feedback from farmers made the invention of Automatic Insect Removal Bin possible. To develop or advance any technology, feedback is highly important. Feedback from users, in my case, farmers impelled me to do basic research, followed by research and development and finally leading to invention.

The Automatic Insect Removal Bin did not happen automatically but was a response to a trigger by a farmer and my ardour to address his practical problem. While I was popularizing the TNAU probe trap technology among the farmers I happened to meet a farmer at a big festival and he asked me a question that was sure enough to jerk me from my complacency and address the pressing issue raised. This, in fact, lay the foundation for my invention of the Automatic Insect Removal Bin that has led to a near revolution in grain storage practices in agriculture.

So what was the question that was fired at me by a humble farmer, which paved the way for the invention? I'm sure it must have triggered your curiosity to know as much as it did mine when I met the farmer as he set the stage before firing the query at me!

His concern was very genuine when he asked, "We are always busy with our farming activities and we may not be able to check the TNAU Probe Trap frequently. Instead why don't you, a scientist, develop a storage bin to store our seeds first without insect infestation at least for a period before sowing them for the next season? Such a bin must be capable of removing insects automatically and farmers should be able to see them."

It was indeed an eye-opener that needed addressing!

As an agricultural scientist it was my responsibility to address the difficulty the farmers were facing and indeed I was answerable to the questions raised by them. I began searching for an answer to this challenge that assailed the farmers.

Soon enough in 1994, I came up with the answer which was published in the renowned national journal, *"Bulletin of grain technology 32 (2):178-181"* and became the prototype for my next invention, i.e., automatic insect removal bin.

In 1996-98, the central government (Indian Council of Agricultural Research) sanctioned 3,10,000 INR under ICAR Ad hoc grant to scale up the automatic insect removal bin for adoption by farmers (in 100kg, 500 kg capacity) for safe storage of seeds/grains.

This is how the TNAU invention, "Insect Removal" bin came to serve our farmers to keep their grain/seeds stored safely, There is no doubt,

'A grain saved is a grain produced;
A seed saved is thousands produced'.

It is not enough for an agricultural scientist to make an invention and rest on his laurels. Once the invention has received the support and the patent for being marketable, that is when the scientist needs to make greater efforts to educate farmers and others about the product and advocate its widespread usage. Hence I knew I could not rest till I reached out to the farmers for whom I had made the invention.

The task was not easy. It was a matter of taking the machine around, demonstrating and seeking help fervently. Moreover, it is the duty of the government to support farmers to adopt and practice the innovative machine.

On his part, the inventor should never get tired. So I was on a continuous run and an endless spree of demonstrating the contraption and its use. Even if a thousand people came to witness your demonstration, but not a single person adopted it, one should never get dejected. That was the lesson I learned and continue to practise.

I continued my advocacy ceaselessly, I never got tired. My morale was boosted during every demonstration effort in the hope that one day the government will come to my help.

And finally, it became a reality in 2003!

Dr. J. L. Harry Jeyaprakash, former Joint Director of Agriculture (JDA) – Vellore, Tamil Nadu, India distributed 150 TNAU bins to small and marginal farmers under Tamil Nadu Government, Drought Prone Area Programme to control insect pests in storage. It was a great initiative to reach out to the farmers with my invention!

TNAU- INSECT REMOVAL BIN USERS

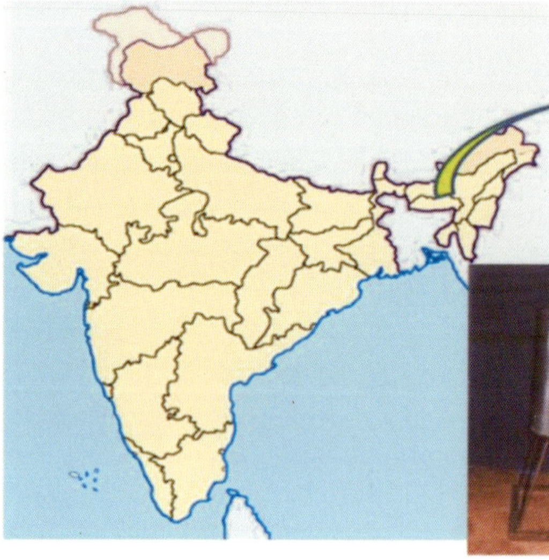

A factory in North-Eastern Zone of India- Manufacturing and marketing the BIN.

Around 5,000 farmers are using the bin for paddy seed storage.

Following the 2003 breakthrough my efforts continued through demonstration and writings in popular publications about the bin, with the hope that the seeds sown by Dr. J.L. Harry Jeyaprakash would yield a rich harvest one day.

Years rolled by, and at last in 2013, the Assam Government launched a massive programme to adopt TNAU automatic insect removal bin for farmers of Assam for paddy seed storage.

This led to the emergence of an entrepreneur in Assam who came forward to manufacture and market the bin for distribution to the farmers of Assam.

This fetched an income of 15, 00,000 INR to Tamil Nadu Agricultural University, thanks to the motivation by the esteemed University.

Being an optimist, I always believe and hope one day all small and marginal farmers in India will have the TNAU innovation viz., insect removal bin for seed / grain storage in all the farm holdings.

Patience and Perseverance are inborn in me and I was confident that this invention would be adopted across the country in good time.

Chapter 3

The Drive behind Obtaining Patents

People used to often ask me if I had patents for my inventions. This question became pertinent especially after I received the Technology Day Invention Award from the Government of India for TNAU insect removal bin in 2002.

The question on patenting came up during the press meet after the award function in Asoka Hotel, New Delhi. Sure enough one of the press persons assembled for the award function queried all dignitaries, "Have all the inventions that have received an award today been patented?"

Hearing this I was shocked since I had not patented any of my inventions so far. Why? The answer was simple. Around that time, there was not much awareness about the importance of patents.

Probably that was the reason the Government of India took a decision to make patenting necessary to apply for an invention award for a product / process after 2002. That patenting is a must is of course a good decision. Though I had already received the award without a patent, I was prompted to make another invention and obtain Indian patent.

And the opportunity materialized soon enough. I got my first patent for an invention which was granted in 2006 for the great institution, Tamil Nadu Agricultural University to which I owe a lot.

At this point the reader would be eager to know what it that guided me toward the path of patented innovation was. This path entailed a long and arduous journey, from India to Canada and back to India to initiate the work on my next important mission.

In 2000, I was deputed to Cereal Research Centre (CRC), Agriculture and Agri. Food Canada, Winnipeg, Manitoba under AHRDP (Agricultural Human Resource Development Programme) scheme of World Bank to get trained in the area of my specialization in stored product insect monitoring under Dr. Paul Fields, Scientist, CRC, Canada.

Indeed, that was a turning point in my research career as I met Dr. S. R. Loschiavo, whom I consider the father of stored grain insect trap. I understood that I was extremely fortunate to have sat on his chair in his room and worked toward completing my mission.

Later on I realized that the journey from India to Canada was providential so that I could be exposed to a situation that would encourage me to make yet another invention and patent it on behalf of Tamil Nadu Agricultural University. Truth be told, I owe a lot to this university that groomed me into the scientist I am today.

I believed that I was blessed by my country and was sent to Canada with a purpose, to serve the farming community on my return to India. The three-month stay in in CRC, Canada had paved way for yet another path-breaking endeavour from me.

The story of patented invention

Dr. Paul Fields, Scientist, CRC, Canada, listened to me and understood my fervent desire for yet another innovation in agricultural science. He took me to CRC library in 2000 Feb-March and got me an important research paper, which sowed the seed for my next TNAU invention.

The research paper was by Quentin, M.E., an entomologist from Michigan State University, and I owe so much to this renowned scientist. Along with his team he had tested the concept of "Tumbling", which he called "bean tumbling", since he wanted to get rid of the bean weevil *Acanthoscelides obtectus* from stored dried beans.

(Quentin, M.E, J.L. Spencer and J.R. Miller. 1991. Bean tumbling as a control measure for the common bean weevil, *Acanthoscelides obtectus* (Say). *Entomol. exp. appl.* 60: 105 - 109).

This paper was a precursor for my dream tuning true.

I learned one important lesson from this paper.

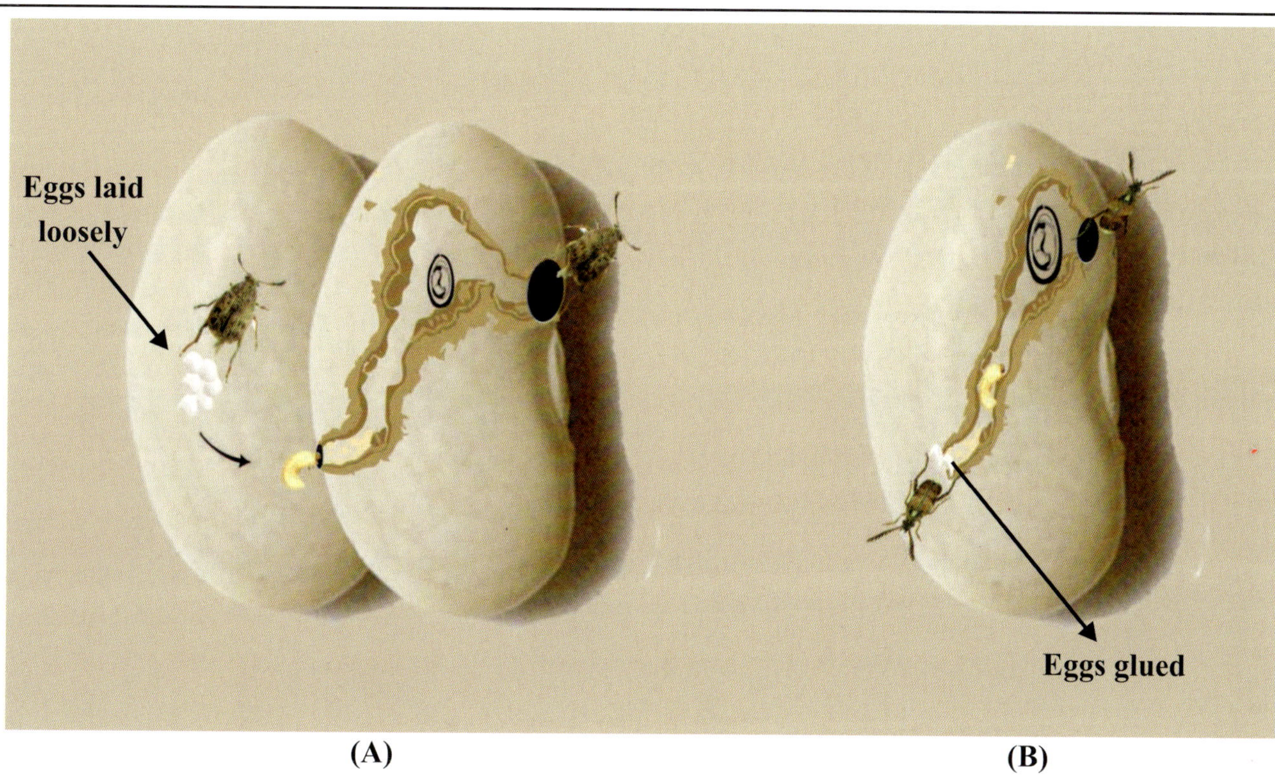

Acanthoscelides obtectus (A) Vs *Callosobruchus* spp. (B)

The insect mentioned by Quentin, M.E is not common in India and has not even occurred in India. The insect lays eggs on the bean seeds which are laid loosely on the seeds. The larvae which hatches from the eggs are mobile and can bore successfully only when they can wedge themselves between abutting surfaces taking nearly 24 hrs or more of nearly continuous scraping to enter the bean seed and settle in (**A**). This behaviour of the insect impelled Quentin, M.E and his colleagues to dislodge the attacking larva by "bean tumbling."

So how did it help me find a solution to grain storage issues assailing Indian farmers?

The insect that afflicts Indian pulse seeds in storage such as green gram, black gram, red gram, beans etc., *Callosobruchus* spp., is indeed a tough guy causing enormous damage during the storage of pulse seed by Indian farmers.

Quentin, M.E did not try to find a solution to prevent the damage of this notorious guy *Callosobruchus* spp., common in India and many developing countries, but just made the following remarks in his paper,

"Tumbling" may be disruptive to insect multiplication by *Callosobruchus* spp., whose larvae bore directly into beans/pulses from eggs glued to the testa, seed coat (**B**).

He stopped with this saying, "some new protocols should be tried," in the hope that a scientist would emerge one day to successfully develop a new protocol to prevent insect damage by *Callosobruchus* spp., I was not aware then that I was preordained to be that scientist!

The research finding by Quentin, M.E (1991) ends here, but for me, this paper was the beginning of a new and exciting journey.

The modus operandi

Soon after my return to India from CRC, Winnipeg, Canada, on 10th April 2000, the very next day I started working in an effort to develop a protocol to avoid the insect *Callosobruchus spp.,* attacking pulse seeds cultivated through arduous toil by Indian farmers and which were the most nutritive components of the Indian diet. ...

I filed a patent with Indian patent office on 10.07.2002 giving details of the Indian Invention as 'A device to remove insect eggs from stored pulse seeds'.

Finally Govt. of India granted patent on 03-02-2006, a momentous event in my life and in turn for TNAU. It was the first patent granted, which led to the birth of a patent era in TNAU history.

I had listened to my inner voice that had murmured,
"Create history, do not repeat history."

I am obligated to the readers to enumerate the technology because I understand now your curiosity must be heightened.

A Device to Remove Insect Eggs from Stored Pulse Seeds (Indian Patent No. 198434)

Pulses are more difficult to store than cereals as these are prone to great damage during storage by pulse beetle *Callosobruchus* spp. The main source of infestation by pulse beetle is through carry over from the field to the stores, which is well known. The present invention is a prototype of a gadget which can successfully crush the eggs of pulse beetle, *Callosobruchus spp.,* which attack stored pulses. The gadget has an outer container and an inner perforated container with a rotating rod fixed with plastic brushes on all sides, which has to rotate one full circumference clockwise and anti-clockwise for 10 minutes, 3 times a day (morning, noon and afternoon). Due to the splashing action of the brush in the rotating rod, the eggs get crushed and thus the seed / grain damage is prevented. The treatment does not affect the germination of the seeds.

Advantages of the invention

1. The device is useful in removing the eggs without affecting the germination.
2. Once the eggs are removed there will be no further build-up of population during the storage of seeds.
3. Removing the eggs laid by the beetles will have a significant impact in arresting the population build up in storage.
4. Farmers are apprehensive of storing pulse seeds fearing damage by pulse beetle during storage. This invention can dispel fear psychosis of the farmers and thus motivate them to store pulse seeds, thereby they benefit immensely.
5. The device is also useful for crushing the eggs, larvae, pupae, and removing the adults (both live and dead) in rice, wheat, dhal, etc.

Once I had developed this insect egg remover, my real work began, that of taking the machine to the farmers and other users. My efforts that began in 2006 culminated in success in 2012.

Around 6 units (prototype, hand operated and power operated) were sold out (see photos next page).

Though 6 is a very small number for an agriculture dominated country like India, this is only the beginning. This is the only available organic, non - chemical, method that is also best suited for community use and the very best for small / marginal farmers / traders of our country about whom we are most concerned. I am confident one day thousands and thousands of machines will come to be used across the length and breadth of our nation helping our farmers, farm traders etc. It will also help us avoid the notorious insects in grain storage, and thereby increase the per capita income of our farmers / traders.

At this juncture, I cannot but help recall the Father of our Nation and the charka (spinning wheel) which he used.

I'm sure there may be questions about any other patent I obtained for another invention. I am happy to disclose that I received yet another patent in 2017- Indian Patent 284727 for a device to monitor Stored Grain Insects in bag stacks. Revolving around this invention is a different but interesting story which the reader may be interested to know.

Capacity of cleaning efficacy 5 kg/hr

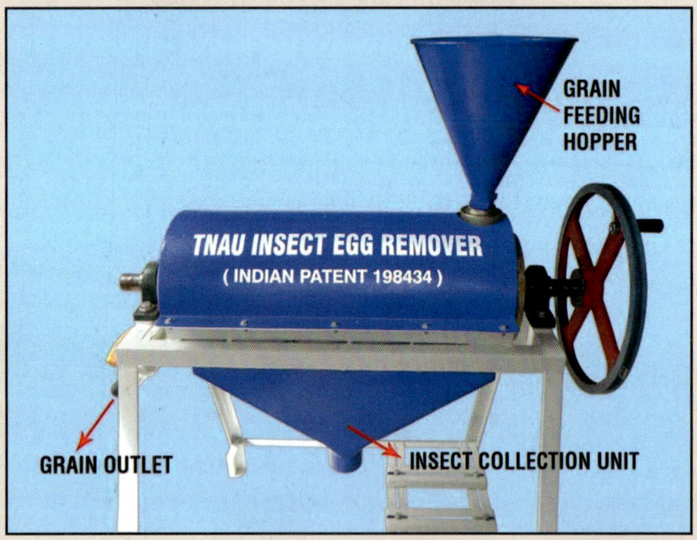

Capacity of cleaning efficacy 50 kg/hr

Capacity of cleaning efficacy 200 kg/hr (1 HP motor)

The Father of our Nation played a major role in getting Independence for our country, which led to the emergence of thousands of scientists in free India, of which I am one. In this picture, I compare the Charka or the spinning wheel used by the Father of our Nation and a symbol of our freedom struggle that led to the independence of India to my TNAU device. While it may not be as powerful as the Charka, it is a great device capable of ushering in a major agricultural revolution in our country - that of Clean Grain (Insect Free) Storage Revolution in India and in other third world countries as well.

Charka - a symbol of swadeshi	**TNAU Egg remover device – Indian Patent Number -198434.(Indian Invention)**
Charka - symbol of Non - Violence.	**TNAU Egg remover device - a symbol of Non - Chemical way to control insects in food grains.**
Charka - simplicity and service to mankind.	**Simple Principle and Service tool for food security mission of India.**
Charka - aimed at removing poverty of village people through supplement of Income by working at home.	**Addition income to farm women who look after grain storage to sell / help others to sell clean, insect free grains by running skill centre with a machine in all villages in India.**
Charka - Symbol in the freedom struggle leading to independence of India.	**Aimed at clean grain (insect free) storage revolution in India.**

Since my last patented invention, my grey cells went on an overdrive and the thinking cap began to tinker on my brain. Intense contemplation led to a question as to why couldn't I develop a simple tool for use by good grain storage warehouse managers, in such a way that not only would the grains be stored easily but they would also be able to know how the notorious insects are distributed in bag stacks stored in warehouses.

An idea clicked and I went on to develop a very simple invention. I applied for Indian Patent on 2008 (1773/CHE/2008) and got it granted in 2017.

A device to monitor insects in bag stacks in warehouse without bait (Indian Patent 284727):

The TNAU - stack probe trap is made of plastic (PVC) which comprises a main hollow tube with a diameter in the range of 1.8 to 2.0 cm with equi-spaced perforation in the range of 1.8 to 2 mm on its upper portion, comprising a bend at one end with a transparent collection unit to collect the insects falling down from the bend, where the other end of main tube is sealed.

TNAU- Stack Probe Trap

Placement of Stack Probe Trap in Turmeric Warehouse

Advantages of the invention:

i. The device is useful in detecting stored grain insects in bag stacks of the food grain warehouses without any damage to sacks.

ii. The device does not require any bait materials to trap insects.

iii. The device is useful in studying the distribution pattern of stored product insects in various layers of bag stacks.

iv. The device will be useful to validate the effect of fumigation by using it immediately after fumigation, in different layers of the fumigated stacks.

v. The device will also be useful at the farm level when farmers store their produce in bags.

Efficacy:

Recent studies established the efficacy of TNAU - Stack Probe Trap in monitoring the pests of stored rice, paddy, wheat flour etc., and also for validating the fumigation in warehouses. The prototype of stack probe trap was tested and found effective in monitoring stored product insects of Siddha and Ayurvedic products. The stack probe trap can detect more precisely the infestation in turmeric compared to normal sampling method during the post fumigation period. It can be used as an effective tool for detection, monitoring of pest and validation of fumigation without damaging the produce and the storage bags.

Commercial plastic model

At present 500 units are being used in the detection of insects in warehouses in India. I am optimistic that every warehouse in India will have this tool, which is an essential tool to detect insects in warehouses which stores > 60 million tonnes of food grains for Public Distribution System in our National Food Security Mission.

Chapter 4

My Cause of Concern for Food Grain Storage Warehouses

Having addressed to certain extent the problems in stored grain insect monitoring at farm and home level.... my eyes fell upon the enormous food grain stored in Indian warehouses for public distribution system which plays a major role in food security of our country.

More than, 60 million tonnes of food grains are stored in bag stacks in Indian warehouses. Now you can imagine the mind-boggling swarms of insects that can invade the grains and products in these warehouses! And this is indeed a matter of great concern because the grains finally make their way to the homes of countless people whose food is prepared using these grains. We owe it to our people to ensure that they get clean and safe food for daily consumption, especially considering that millions of children and young adults consume these grains at the growing stage of their lives.

I am not worried about the statistics of food grain losses caused by insects but, my goal is to develop a suitable technological invention to minimize storage loss in the warehouses. I knew my thinking was in the right direction but it materialized only in the year 2018. So what is the story behind this new endeavour? Truth be told, the story had its beginning way back in 1992.

The out-of-the-box thinking that generated in my mind drew my eyes to view the Indian warehouse with a new approach. This is when I started working on my project of developing a technology to mass trap insects in warehouses.

The enlightenment for this new endeavour came from the 'Light Trap', TNAU UV light trap, which basically stemmed from the well-known literature on how light attracts insects. For the first time in India I did a 'Practical study on the use of UV Light Trap in Warehouses' (My work carried out in 1992-93 was published in *International Journal of Pest Management, 1994, 40 (2): 148 -152).*

Finding this effective, Tamil Nadu Government released the technology of "UV light trap for warehouses" in 1992.

I am sure most of you would have an idea what my next move would be considering that you know how restless I become until the technology that I develop is put to use by the end users. In this case the end users were food grain storage warehouses in our country.

I wrote many articles in eminent publications to campaign for the new technology with an intention to spread its use.

Significant outcomes of technology transfer by Indian Grain Storage Working Group

- Indian Council of Agricultural Research (ICAR) formed the Indian Grain Storage Working Group Vide D.O./IGSWG/2015, Dated, the 26th June-2015 with Dr. K. Alagusundaram, Deputy Director General(Agri. Engg.) as Chairman, IGSWG
- Under the technical activities of IGSWG, TNAU UV Light was tested by Food Corporation of India (FCI).

THE 'TRAP' EXPERIMENTS

@ Food Storage of Depot, Bicavolu, East Godavari District, Andhra Pradesh

- The UV light traps were found to be effective in reducing the insect population inside the sheds.
- They were also effective in godowns situated near residential areas by preventing the entry of flying insects during the night.

This led to a large scale adoption of UV Light Trap by FCI as well as by many private Warehouses across India. (around 1000 users)

- Large scale production by M/s Melwin Engineering, Coimbatore, Tamil Nadu, India, who has been given License by Tamil Nadu Agricultural University.
- Promotion of a small scale Enterpreneur

Significance :

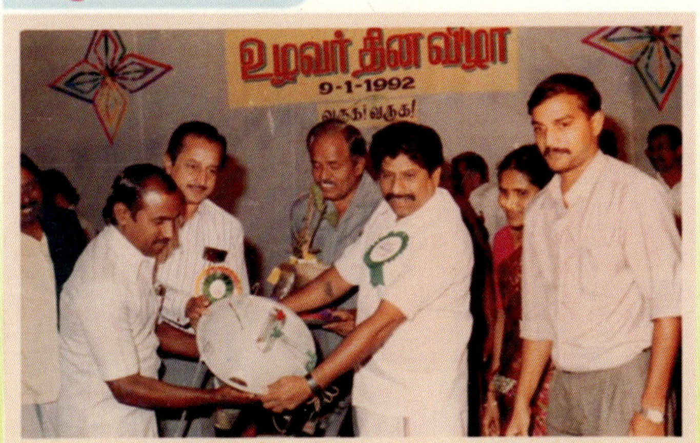

TNAU UV Light Trap release in 1992

Though TNAU Light Trap was Introduced in 1992, the technology was transferred on a Large scale to FCI only after the formation of Indian Grain Storage Working Group in 2015, which formed strategies for effective Grain Storage Management.

The qualities of patience and perseverance inherent in me came to the fore as I strongly believe these traits eventually trigger the inventor to tread the success trajectory.

My belief was justified in 2002 when M/s Madhaus Pharmaceuticals, Goa adopted the TNAU UV light trap (around 12 units) technology for the management of Cigarette beetle, *Lasioderma serricorne* in herbal tea warehouses.

Then after 2008 there was a lull.

My efforts continued till I could see that the Food Corporation of India, which stores more than 60 million tonnes of food grains for public distribution system, used this UV light trap technology for mass trapping insects in warehouses.

My belief, "Create history," coupled with faith in God as well as in myself saw me attain yet another significant success in my career as a scientist who wanted to contribute to the country, to the University that had supported him and to the countless people of our country.

My efforts finally resulted in making TNAU - UV light trap technology an integral part of insect management in food grain warehouses in India. This is a great success and victory for an Indian innovation and a reward for untiring efforts and relentless work which finally yielded a sense of satisfaction.

Chapter 5

How my Technology Influenced Graduate Students

When we have a vision to make a positive contribution to the society, the nobility of the vision itself leads to its success. My endeavours were fervently directed towards introducing the TNAU gadgets for stored grain insect in the education system so as to benefit the students of our country as well as those across the globe.

When we have a clear vision, we can see ways to achieve the same. In my journey, time and again faith, hope and the desire to transform the fruits of my labour to those that help and make a difference in the lives of many have stood me in good stead, and unveiled the path of success, despite time lapses or impediments, which were circumvented with positivism. Providentially, Canada seems to have played an important role in fructifying my dreams and I am must obliged to the country as well as its great scientists.

In 2005 I was deputed to undergo a training for six months under TNAU – McGill Canadian International Development Agency Project on Food Security in South India at McGill University, Quebec, Canada. There I got the help of Dr. Venkatesh Sosle and Dr. G.S. Vijaya Raghavan, Professor, McGill University to realise my vision of taking the TNAU gadgets to students across my country. They helped me fund my idea on developing a “Tool kit” for teaching and training.

The TNAU kit

The TNAU stored grain insect management kit along with a CD on the various improved technologies (gadgets) developed by me in the Tamil Nadu Agricultural University served as a 'tool' for the outreach activity to train the students, scientists, extension personnel, households, entrepreneurs and warehouse quality control managers.

Description of TNAU Stored Product Insect Pest Management Kit

Food grains are stored for varying periods to ensure proper and balanced public distribution throughout the year. Among the biotic and abiotic factors which affect grain / seed in storage, insect plays a major role in the deterioration of grains / seeds causing both qualitative and quantitative loss. Often the presence of insects in store houses are felt only when the insects are found hovering and flying around, by which time there is enormous loss and population build-up of insects. Hence, timely detection of the stored grain insects will help prevent heavy loss.

TNAU - Kit Technology spread Across the country

The TNAU Kit was launched by Tamil Nadu Agriculture University, Coimbatore. on 15th September, 2005.

TNAU - Kit

I Traveled with Kit to Popularize

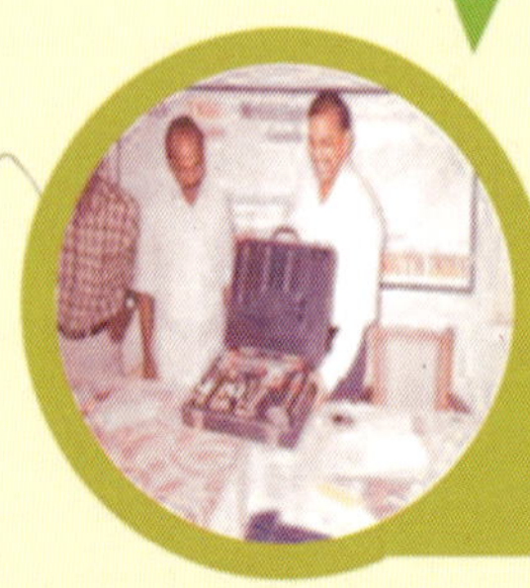

Palladam, Tamil Nadu.

Madurai, Tamil Nadu.

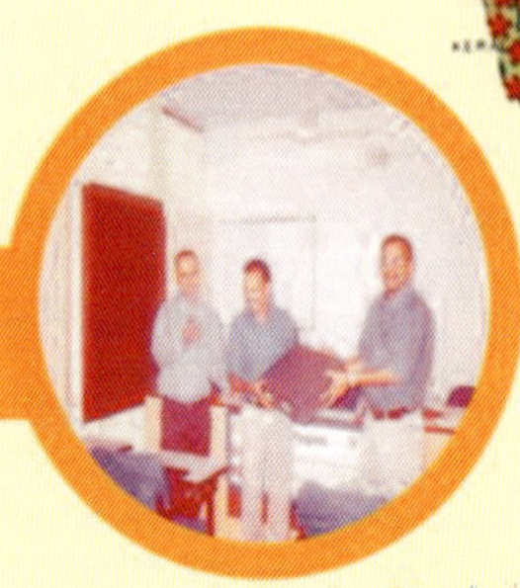

Trichur, Kerala.

Anantapur, Andra Pradesh

Through my efforts TNAU emerged as one of the pioneering institutes in India in the development of detection devices for stored grain insects. These devices exploit the wandering behaviour of the insects and help in timely detection of insects in stored produce leading to timely control. These include TNAU probe trap, TNAU pit fall trap, two-in-one model trap, indicator device, Automatic insect removal bin, and UV-light trap technology, a device to remove Insect Eggs from Stored Pulse Seeds (Indian Patent No. 198434) and a device to monitor insects in bag stacks in warehouse without bait (Indian Patent 284727). These devices have been widely used in many places and have received State and National recognitions.

Hence, I developed a "KIT" named as **TNAU-Stored Grain Insect Pest Management Kit** containing prototypes of all the devices along with a CD-ROM about the devices and how to use them. This kit immensely helped in popularizing the technologies across the country. The kit proved to be an ideal hands-on training tool in Education, Extension centres (KVK, Plant clinics, save grain centres) and also in private warehousing. This TNAU kit is the first of its kind in the world.

By now you might have surmised that, I never rest after developing a technology, I did extensive demonstration in many regions of our country on the kit technology for educating college students.

The TNAU Kit was launched by Tamil Nadu Agricultural University, Coimbatore on September 15, 2005.

The Photo Show depicts the spread of TNAU-Kit Technology.

Chapter 6

Kindling Scientific Temper among School Children

"If we are to teach real peace in this world, we shall have to begin with the children"

-- *Mahatma Gandhi*

For me, realising this dream of the father of our nation adds a lot of meaning to my life. I went beyond the realm of college students and my outreach finally encompassed school children too. It was a very gratifying experience for me.

To be able to take the gadgets to school seemed a distant dream. But it did materialize, with the grace of God and the support of the government of Tamil Nadu. I consider this to be a miracle in my scientific career.

Impact on School Children

My traps kindled the scientific temper in school children in Tamil Nadu

(This was possible only by extensive popularization agenda done by me)

Besides being useful to the farmers, households and warehouse managers, my TNAU Insect trap technologies paved way for kindling scientific temper in school children.

THE HINDU, Friday, January 17, 2003

Making cost-effective pest traps

IN THEIR eagerness to achieve 100 per cent result, most schools usually end up churning out children with a high score card but with little holistic development. Parents are usually blamed for exerting pressure on their wards to score high marks to facilitate easy entry into professional courses.

In the process, scant regard is paid to extracurricular activities. However, in the present environment, in addition to academics, schools should also strive to bring out the latent talent among students and kindle their scientific temper.

In this regard, anything that's creative and innovative usually tends to grab attention. Students of the Senthil Matriculation School in Dharmapuri recently won accolades for a model based on the theme 'Caring and storing food grains from pests'.

It also fetched them a medal at the 10th National Children's Science Congress-2002.

But the win wasn't a cakewalk for this batch of sixth standard kids - N. R. Badrinath, M. Hari Vignesh, P. Rajapathy, S. Deepan and R. P. Harish. And the inspiration for the project came from none other than the science teacher K. Kalavathi, who had earlier exhibited a model of 'Mohan trap', designed by a TNAU scientist, Mohan.

The trap priced at Rs.45 to Rs. 60 works by luring insects into a perforated tube like structure with a cap at the end.

The pests then enter the tube and gets collected in the cap. The trap is being marketed after being subjected thorough several tests. Based on this concept, the children developed their own model with special focus on making it economical for the rural masses.

The student team, with assistance from their teacher, struck upon the idea of using throwaway plastic soft-drink bottles for the making the trap.

By making holes in the bottle, just big enough for the pests to enter, the team sliced the bottle into two. Placing a funnel near the mouth of the bottle enabled the pests to be directed into the bottle cap, which automatically serves as the collection point.

The students then joined the bottle using a tape and tested it. However, with this system entailing expenditure towards the purchase of funnels, the students created funnels using thick paper.

As it turned out to be successful, the students have proceeded to create a cost-free pest trap.

It was called 'cole trap' based on a common type of pests found in grains called rice weevil (coleopetra). Ms. Kalavathi said, "When the idea of making a pest trap was mooted to the children, there was a deluge of ideas on means to make it more cost-effective, hence the project was immediately selected for the competition. At the final stage, suggestions by the students of replacing the funnel that cost Rs.2, with mere thick paper, made the trap absolutely free."

For the students it had been a tough task reaching the Science Congress held at Mysore.

Initially, the project on being approved at the district level was sent to the State level competition held at Tirupur, in which 55 schools participated.

On being selected there, the students reached the finals where they won a prize among the 610 entries on display.

Badrinath, who had made the presentation at Mysore, said, "The teacher, while discussing about post harvest technologies, had discussed about earlier practice of placing a mixture of Vasambu and Neem. She had also demonstrated the present day practice of using these traps, which prompted us in finding out ways to make it as cheap as possible. But we finally ended up in creating something that could be made totally free of cost."

The organisers in their certification have made the following commendation about the project "we find your choice of project great. A good problem chosen is half the job done! You have shown a remarkable ability to succinctly present the work. The logical structure, choice of aids and clarity of presentation is superb. It's a brilliant piece of work overall."

By D. Sivarajan in Dharmapuri

Tapping hidden talent

A. **Seventh Standard Student of Chinmaya Vidyalaya, Trichy, Tamil Nadu Mr. M. Abhilash won 2 Gold Medals in the INTEL - IRIS completion and a WIPO Award** for young inventor by making TNAU traps using waste materials (Bisleri bottles).

B. **Students of the Senthil Matriculation School in Dharmapuri, Tamil Nadu recently** won a Medal at 10th National Children's Science Congress - 2002 **by developing their own model of insect trap from the inspiration they got on seeing my probe trap model.**

C. **URC Palaniammal Matric Higher Secondary School, Sengodampalayam, Erode, Tamil Nadu has won a prize and INR 1000/cash award for the model** "Protection of Food Grains" **in the exhibition cum competition,** "Genesis 2003" **organized by P.K.R. Arts College for Women, Gobichettipalayam.**

D. **Mr. R. Vignesh of Padma Sheshadri Balamandir Higher Secondary School, Chennai, Tamil Nadu won the best project award for making TNAU traps in plastic waste materials during** SCHOOL ANNUAL MEET 2009**.**

The Tamil Nadu Government, impressed by the creative talent of the school students appointed me as the **Chairman of the Committee to revamp and revitalize the Agricultural School Education in Tamil Nadu during 2010. I felt this was the greatest achievement in my life!**

Education is very important in shaping the future of the student community.

The new 11th Standard Agricultural Practices I Book has a chapter on the Importance of Post-Harvest Technology in which the TNAU probe trap is highlighted. This stood testimony to the words of Robert John Meehan, 'Teachers who love teaching, teach children to love learning'.

As the Chairman I made significant reforms in Agricultural education in Tamil Nadu schools. The old curriculum structure which was in practice in Tamil Nadu for the last 20 years was revamped and a common text book Agricultural Practices I and Agricultural Practices II (for 11 and 12th standard schools) was introduced in the year 2010-11, 2011-12 respectively. For the First time in the curriculum, Practical Guide was introduced for 11th and 12th standards in Tamil Nadu.

Significant reforms

Earlier there were many subjects like Crop Management, Crop Protection, Vegetable Cultivation, Agricultural Chemicals, Poultry, Animal Husbandry, Fisheries and Marketing. Schools were free to choose any one of the books on the subject given above for 11th and 12th standards. This was in practice for several years.

Disadvantage

Students from the above stipulated curriculum studied only one subject e.g.: A students who completes 11th and 12th standard in "Crop Protection" knows only crop protection without any knowledge of crop science. This was similar for those who studied other subjects too. This led to a great problem when they got admission in Diploma or Degree courses after 12th standard in Agriculture. Hence I made a major change in this concept.

Further, I also incorporated changes to make the school curriculum more practical than before.

Major Reforms

Students undergoing 11th and 12th standard in vocational stream Agriculture will now have two books:

11th Standard Agricultural Practices I. This gives a broad, basic knowledge on agriculture, including Animal Husbandry, Fisheries Marketing, and Farm Machinery etc.

12th Standard - Agricultural Practices II. This gives a detailed knowledge on Important Crops of Tamil Nadu, Cultivation and Management, Animal Husbandry Management etc.

For the First Time

- Practical guide for 11th standard and
- Practical guide for 12th standard were introduced.

This curriculum helps students acquire skills in practicing agriculture so that they can get employment in agro-based industries after their school education.

E.g.: Seed industry, Honeybee rearing, Cut flower making, Fish culturing, Animal Husbandry management etc. They can even be self-employed.

The book cover page in my regional language Tamil 2010

© தமிழ்நாடு அரசு
முதல் பதிப்பு – 2010

குழுத்தலைவர்

முனைவர். ச. மோகன்,
பேராசிரியர் (பூச்சியியல் துறை),
தமிழ்நாடு வேளாண்மைப் பல்கலைக்கழகம்,
கோயம்புத்தூர் – 641 003.

நூல் ஆசிரியர்கள்

முனைவர் த. வசந்தி,
பேராசிரியர் (மண்ணியல் மற்றும் வேளாண் வேதியியல் துறை),
தமிழ்நாடு வேளாண்மைப் பல்கலைக்கழகம்,
கோயம்புத்தூர் – 641 003.

அ. ஆனந்தகலைச்செல்வி,
தொழிற்கல்வி ஆசிரியை (வேளாண்மை),
அரசு ஆண்கள் மேல்நிலைப்பள்ளி,
தொண்டாமுத்தூர் – 641 109.

பெ. திருமால்காந்தி,
தொழிற்கல்வி ஆசிரியர் (வேளாண்மை),
அரசு மேல்நிலைப்பள்ளி,
புஜங்கனூர் – 641 113.

இரா. மலர்விழி,
தொழிற்கல்வி ஆசிரியை (வேளாண்மை),
அரசு மேல்நிலைப்பள்ளி,
காரமடை – 641 104.

ச. ஹில்டா,
தொழிற்கல்வி ஆசிரியை (வேளாண்மை),
அரசு ஆண்கள் மேல்நிலைப்பள்ளி,
பல்லடம் – 641 664.

பாடங்கள் தயாரிப்பு : தமிழ்நாடு அரசுக்காகப்
பள்ளிக் கல்வி இயக்ககம், தமிழ்நாடு

இந்நூல் 60 ஜி.எஸ்.எம்.தாளில் அச்சிடப்பட்டுள்ளது

வெப் ஆப்செட் முறையில் அச்சிட்டோர்

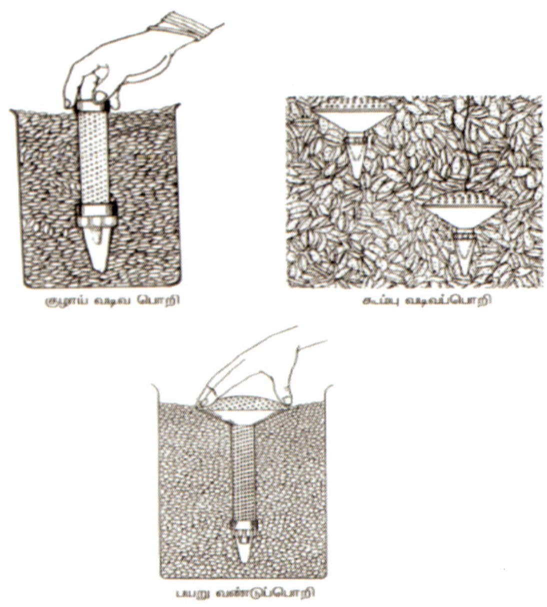

படம் 53. தானிய சேமிப்புப் பொறிகள்

193

Advantage of present Syllabus

The greatest advantage of the present syllabus is that students who pass out after 2012 June would have more practical knowledge on

1. Agriculture
2. Veterinary Science
3. Horticulture
4. Fishery
5. Forestry
6. Agricultural Engineering

Whereas, before the introduction of this syllabus, students did not have these advantages. Equipped with this knowledge of agriculture and given the competition, our rural students can perform well in the Degree / Diploma Programmes and would be comparable to other science students of 11th and 12th standards.

This is a revolutionary step in Rural Agricultural Education I made as the Chairman of the Committee to Revamp and Revitalize the Agricultural School Education in Tamil Nadu.

Through this new syllabus a great door has been opened for our rural students who mainly study Agriculture in their +1 and +2 level to become Graduates in AGRICULTURAL SCIENCES of our country besides becoming Graduates in VETERINARY SCIENCE. This I feel the greatest achievement in my life.

I fervently hope and believe that my journey to kindle scientific temper among school children will continue ceaselessly.

Chapter 7

Scientific Pursuit is Endless

What Next ...? Go forward... Way ahead...

**My Efforts will Continue in Popularizing
Non - Chemical Food Grain Saving Technologies
Across the World.**

Chapter 8

References

1. Mohan, S., M. Gopalan, P.C. Sundara Babu and V.V. Sreenarayanan. 1994. Practical studies on the use of light trap and bait trap for management of ***Rhyzopertha dominica*** in rice warehouses. **International Journal of Pest Management.., 40 (2): 148-152.**

2. Mohan, S. 1994. A storage container model for automatic removal of stored product insects from grains. **Bulletin of Grain Technology, 32 (2): 178-181.**

3. Mohan, S. 2000. Insect removal bin could help farmers in developing countries. **Resource, 7(9): 6 (Journal of American Association of Agrl. Engineers).**

4. Mohan, S., C.T. Devadass and D. Mahendran. 2001. New devices for pulse beetle management. **Plant Protection Bulletin, 53 (3&4): 27-32.**

5. Mohan, S. and Paul Fields. 2002. A simple technique to assess compounds that are repellent (or) attractive to stored product insects. **Journal of Stored Product Research, 38: 23-31.**

6. Mohan, S. 2002. Light traps. In: **Encyclopaedia of Pest Management. p. 455-459. Marcel Dekker Inc. New York.**

7. Mohan, S., S.S. Sivakumar, S.R. Venkatesh and G.S.V. Ragavan. 2006. Simple technique to increase the sensitivity of probe traps in detecting *Cryptolestesferrugineus* in stored wheat. **Canadian Entomologists 2006. 138 (5): 733-735.**

8. Mohan, S., S.S. Sivakumar, S.R. Venkatesh and G.S.V. Ragavan. 2007. Penetration of polyethylene sheets coated with protein enriched pea flour solution by two stored product insects. **Journal of Stored Product Research 2007. 43: 202-204.**

9. S. Mohan. 2007. Eco friendly post-harvest technologies for management of stored grain insects. **Green farming 1 (2): 45-47.**

10. Mohan, S. and Zadda Kavitharaghavan, 2008. Systematic feedback studies on TNAU Probe trap in Pallapalayam, Palladam in Coimbatore, Tamil Nadu. **Pestology, XXXII, (6): 49-51.**

11. Mohan, S., S. Sivakumar, Z. Kavitharaghavan, S. Venkatesh and G.S.V. Ragavar. 2008. A new trap model to increase the trapping of *Cryptolestesferrugineus* (Colleoptera : Laemophloeidae) in wheat filled containers. **Madras Agricultural Journal 95 (7-12): 390-393.**

12. Mohan, S. 2011. A device to remove adult insects and crush their eggs in stored grains. **Pestology *Vol. XXXV No.9*, 29-30.**

13. Mohan, S and A. Rajesh. 2016. Use of light traps in a phosphine resistance management strategy for *Tribolium castaneum* (herbst) in Indian grain storage warehouses. **In**: **Proceedings of The 10th International Conference on Controlled Atmosphere and Fumigation in Stored Products, New Delhi, 7 – 11, Nov, 2016. Ref. No.7 0-4.**